C++

FOR BEGINNERS

The Ultimate Guide To Learn C++
Programming Step-by-Step

By Tom Clark

Introduction

C++ is a for the most part significant programming language proposed to make programming more charming for the authentic software engineer. With the exception of minor subtleties, C++ is a superset of the C programming language. In spite of the working environments given by C, C++ gives adaptable and fruitful work environments to depicting new sorts. A softwareengineer can disperse application into sensible pieces by depicting new sorts that enthusiastically match the musings of the application. This technique for program construction is routinely called information thought. Objects of some client depicted sorts contain type data. Such things can be utilized well and securely in settings in which their sortcan't be hinder mined at arrange time. Adventures utilizing objects of such sorts are constantly called object based. Right when utilized well, these situations accomplish more confined, more plainly obvious, and simpler to really focus on programs.

The significant idea in C++ is class. A class is a client depicted sort. Classes give up informationcovering, ensured instatement of information, seen sort change for client depicted sorts, dynamic shaping, client-controlled memory the heads, and instruments for over-upsetting chiefs. C++ gives much better work environments to type checking and for passing on separation than C does. It's like way contains upgrades that are not immediate identified with classes, including significant constants, inline substitution of cutoff points, default work clashes, over-inconvenience work names, free store the boss's administrators, and a reference type. C++ holds C's capacity to manage the central objects of the equipment (bits, bytes, words, addresses, and so on) This permits the client portrayed sorts to be executed with an awesome level of ability. C++ is an intermediate level language, as it contains a confirmation of both unquestionable level and low-level language features. C++ is a free form,statically type, multiparadigm, compiled general purpose language. C++ is an Object-Oriented Programming language at some points isn't totally Object Oriented. Its features like Friend and Virtual, contradict a bit of the essential OOPS features. Hence you can call it both an intermediate programming language as well as object-oriented programming language. C++ and its standard libraries are expected for portability. The current execution will run on most systems that help C. C libraries can be used from a C++ program, and most instruments that

Chapter 1

1.1 Description

C++, as we know in general is a wing to C language and was made by Bjarne Stroustrup at belllabs. C++ was intended to give Simula's working environments to program relationship close by C's ability and flexibility for structures programming. It was needed to give that to credibletasks inside a colossal section of a time of the thought.

The objective was really simple at thattime it didn't involve any sort of innovation and was quite a compromise on the flexibility andefficiency of the language. While an unassuming degree of progress emerged all through theextended length, efficiency and flexibility have been kept up without deal. While, the destinations for C++ have been refined, clarified, and made all the more express all through the long haul, C++ as used today clearly reflects its exceptional focuses. Most effort has been utilized on the early years because the arrangement decisions taken early chose the further improvement of the language. It is also less complex to keep an unquestionable perspective when one has had various years to see the aftereffects of decisions.

In this short book we will talk about how C++ language emerged over the time and what type of programming updates have occurred over the years.

Chapter 2

2.1 First Programming Language (Simula)

2.1.1 History

The essential object-oriented programming language was made during the 1960s at the Norwegian Registering Center in Oslo, by two Norwegian PC specialists—Ole-Johan Dahl (1931-2002) and Kristen Nygaard (1926-2002).

Kristen Nygaard, a MS in number juggling at the College of Oslo, started creating PC entertainment programs in 1957. He was searching for a better strategy than depict the heterogeneity and the movement of a system. To go further with his contemplations on an ordinary scripting language for depicting a system, Nygaard comprehended that he requiredsomeone with more PC programming capacities than he had, thusly he arrived at Ole-Johan Dahl, in like manner a MS in math and one of the Norway's chief PC analyst, who obliged himin January 1962.

In 1966 the English PC scientist Tony Hoare introduced record class create, which Dahl and Nygaard connected with prefixing and various features to meet their necessities for another summarized measure thought. The essential customary importance of Simula 67 appeared inMay 1967. In June 1967 a gathering was held to standardize the language and start differentexecutions. Dahl proposed to unite the sort and the class thought. This provoked real discussions, and the suggestion was excused by the board. SIMULA 67 was formally standardized on the chief fulfilling of the SIMULA Rules Gathering in February 1968.

Simula 6 contained huge quantities of the thoughts that are presently open in standard Object-oriented like Java, C++, and C#.

2.2 C++ Language evolution

The C++ programming language has a bunch of encounters getting back to 1979, when Bjarne Stroustrup was handling position for his Ph.D. Proposition. One of the vernaculars Stroustrupgot the opportunity to work with was a language called Simula, which as the name recommends is a language essentially proposed for reenactments. The Simula 67 language - which was the variety that Stroustrup worked with - is seen as the vital language to help the article masterminded programming perspective. Stroustrup found that this perspective was outstandingly important for programming improvement, in any case the Simula language was nonsensically postponed for sensible use.

2.3 C with classes

"C with Classes" was the earlier version of C++. C++ got evolved through it. The main purposeof C with classes was to add classes in to the language. This work happened in between 1979-1983. This work determines the shape of C++.

The work, on what finally became C++, started with an undertaking to analyze the UNIX partto choose the amount it might be passed on over an association of laptops related by an area.This work started in April of 1979 in the Figuring Science Exploration Focus of bell Labs in Murray Slope, New Jersey, the started forward. Two subproblems after a short time emerged: how to examine the association traffic that would result from the piece movement and how to modularize the part. Both required a way to deal with impart the module plan of a perplexing structure and the correspondence illustration of the modules. This was overall such an issue that had become concluded never to attack again without real instruments. Consequently, the development of a genuine contraption according to the models that wereoutlined in Cambridge.

During the April to October period the advancement from thinking about a "gadget" to contemplating a "language" had occurred, yet C with Classes was at this point considered ona very basic level as a growth to C for conveying estimated quality and synchronization. A fundamental decision had been made, notwithstanding. Notwithstanding the way that help of concurrence and Simula-style reenactments was a fundamental place of C with Classes, thelanguage contained no locals for imparting concurrence; rather, a mix of heritage (class levelsof leadership) and the ability to portray class part works with outstanding ramifications saw by the preprocessor was used to create the library that maintained the ideal styles of synchronization. Mercifully note that "styles" is plural. I considered it basic, as I really do, that more than one thought of concurrence should be expressible in the language. This decision has been reconfirmed more than once by me and my partners, by other C++ customers, and by the C++ standards warning gathering. There are various applications for which maintain for concurrence is principal, yet there is no one winning model for synchronization maintain; hence when sponsorship is required it should be given through a library or a particular explanation increase with the objective that a particular sort of concurrence maintain doesn't hinder various constructions.

As such, the language gave general instruments to figuring everything out programs instead of help for express application zones. This was what made C with Classes, and later C++, an extensively valuable language rather than a C variety with expansions to help specific applications. A short time later, the choice between offering assistance for explicit applications or general thought frameworks, has come up more than once. Each time the decision has been to improve the reflection segments.

An early depiction of C with Classes was conveyed as a bell Labs particular report in April 1980[Stroustrup 1980a], and later in SIGPLAN Takes note. The SIGPLAN paper was in April 1982, followed by a more point by point Chime Labs specific report, "Adding Classes to the C Language: An Activity in Language Development" [Stroustrup 1982], that was thusly appropriated in Programming: Practice and Experience. These papers set a real model by portraying simply features that were totally executed and had been used. This was accordingto a long-standing act of bell Labs Figuring Science Exploration Center; that plan has been changed exactly where more openness about the inevitable destiny of C++ got expected to ensure a free and open conversation over the headway of C++ among its various non-AT&T customers.

C with Classes was explicitly expected to allow better relationship of activities; "estimation" was seen as an issue tended to by C. The expressing point was to facilitate with C in regard to run-time, code minimization, and data diminutiveness. In reality: someone once showed a three percent purposeful decrease in commonly run-time efficiency compared with C. This was seen as unsuitable and the overhead promptly disposed of. In like manner, to ensure design likeness with C and therefore avoid space overheads, no "housekeeping data" was placed in class objects.

Another critical concern was to avoid impediments on the space where C with Classes could be used. The ideal-- which was cultivated - was that C with Classes could be used for whateverC could be used for. This proposed that just as organizing with C in efficiency, C with Classes couldn't offer advantages to the impediment of disposing of "dangerous" or "revolting" features of C. This discernment/standard should be repeated habitually to people (rare C withClasses customers) who required C with Classes made safer by growing static sort checking according to early Pascal. The elective technique for giving "security," embeddings run-time checks for each and every dangerous action, was (and is) considered reasonable for investigating conditions, at this point the language couldn't guarantee such checks without leaving C with a huge advantage in run-reality viability. Hence, such checks were not given for C Classes, anyway C++ conditions exist that give such checks to examining. Besides, customers can, and do, implant run-time checks (revelations [Stroustrup 1991]) where required and sensible.

C allows low-level errands, for instance, bit control and picking between different sizes of entire numbers. There are furthermore workplaces, as unequivocal unchecked sort changes, for deliberately breaking the sort structure, C with Classes, and later C++, follow this path byholding the low-level and perilous features of C. Instead of C, C++ intentionally discards the need to use such features beside where they are basic and performs hazardous undertakings exactly at the explicit request of the programmer. As every programmer have different stylesand ways to write a program and most certainly every problem have many ways to solve it so the language should be developed in a way so as to encourage a programmer to write his style out with only essentials of language to follow.

Features that were included in 1980:

1. Friend classes
2. Derived classes
3. Constructors and destructors

4. Public/private access control
5. Type checking and conversion of function arguments
6. Friend classes
7. Call and return functions (Section 15.2.4.8)

The features that they included into the language during 1981

8. Overloading of assignment operator
9. Default argument
10. Inline function

Since a preprocessor was used for the execution of C with Classes, simply new features, thatis, features not present in C, ought to have been depicted and the full power of C was direct open to customers. Both of these perspectives were esteemed by then. Having C as a subset fundamentally reduced the assistance and documentation work required. This was generallyhuge because for a serious drawn-out period of time I did the total of the C with Classes andlater C++ documentation and sponsorship just as doing the experimentation, plan, andexecution. Having all C features open additionally ensured that no requirements introduced through inclination or nonappearance of foresight on my part would prevent a customer from getting features viably open in C. Typically, convey ability to machines supporting C was ensured. From the start, C with Classes was done and used on a DEC PDP/11, yet soon it wasported to machines, for instance, DEC, VAX, and Motorola 68000-based machines. C with Classes was at this point seen as a vernacular of C. In addition, classes were insinuated as "A Theoretical Information Type Office for the C Language" [Stroustrup 1980a]. Support for object-orchestrated composing PC programs was not ensured until the course of action of virtual limits in C++ in 1983 [Stroustrup 1984a].

Key design decisions of that time:

1. C with Classes follows Simula in permitting the designer to decide types from which factors(objects) can be made, rather than, say, the Modula approach of showing a module as a grouping of articles and limits. In C with Classes (as in C++), a class is a sort, this is a fundamental thought in C++.

2. The depiction of objects of the customer portrayed sort is fundamental for the class introduction. This has broad consequences. For example, it suggests that certifiable area components can be done without the use of free store (load store, dynamic store) or refuse grouping. It moreover infers that a limit ought to be recompiled; the depiction of a thing it uses clearly is changed.

3. Gather time access control is used to bind permission to the depiction. As is normally done,only the limits referred to in the class confirmation can use names of class people. People

(when in doubt work people) showed in the public interface, the assertions after the public: name, can be used by other code.

4. The full kind (tallying both the return type and the conflict kinds) of a limit, is demonstrated for work people. Static (request time) type checking relies upon this benevolent detail. This changed from C by then, where work conflict types were neither decided in interfaces nor checked in calls.

5. Limit definitions are usually designated "elsewhere" to make a class more like an interfacespecific than a lexical instrument for figuring everything out source code. This gathers that extraordinary. This implies that separate accumulation for class part abilities and their clientsis simple and the linker period generally utilized for C is adequate to help C++.

6. The function new () is a constructor, a function with an uncommon which intends to the compiler. Such functions outfitted guarantees around directions. In this model, the assuranceis that the constructor, perceived actually confusingly as another trademark, on the time is destined to be alluded to as to introduce each object of its class before the primary utilizationof the object.

7. Both pointers and non-pointer types are provided (as in every C and Simula).

Much of the further improvements of C with classes and C++ may be viewed as investigating the consequences of these design decisions, misusing their proper components, and makingup for the issues coming about because of their awful features. Many, besides in no way all, of the ramifications of those layout picks were perceived at that point; Stroustrup [1980a] is dated April 3, 1980. This level endeavors to make clear what changed into comprehended atthe time and bring suggestions to segments clarifying later effects and acknowledge.

2.4 Run-Time Efficiency

The initial version of C with classes did not provide inline functions to take in addition benefit of the supply the illustration. Inline functions had been quickly provided, even though. The general purpose for the creation of inline functions became fear that the cost of crossing a protection barrier would possibly reason people to chorus from the use of commands to coverrepresentation. Specially, Stroustrup [1982] observes that human beings had made records contributors public to avoid the function name overhead incurred with the useful resource ofa constructor for simple schooling wherein most effective one or two assignments are needed for initialization. The instantaneous reason for the inclusion of inline capabilities into C with classes turned into a venture that could not control to pay for feature call overhead for a fewcommands involved in actual-time processing.

Over time, issues along those strains grew into the C++

"precept" that it became not so sufficient to offer a function, it had to be supplied in a less pricey form. Maximum really, "low fee" grow to be seen as that means "low-priced on hardware were common among developers" in area of "much less luxurious to researchers with excessive-quit device," or "low-cost in a couple of years while hardware can be less expensive." C with classes was continuously considered as a few things to be used now or subsequent month in choice as studies project to supply something in multiple years, therefore. In lining changed into taken into consideration critical for the application of education and, therefore, the difficultyemerge as more a way to offer it than whether or how not to offer it. Two arguments receivedthe day for the belief of getting the programmer pick out which functions the compiler needto attempt to inline.

The compiler best knows quality if it's been programmed to inline and it has a notion of time/space optimization that agrees with mine. The alternative languages become that most effective "the subsequent release" could actually inline and it might achieve this consistent with an inner good judgment that a programmer couldn't successfully manipulate. To make matters worse, C (and therefore, C with lessons and later C++) has authentic separate compilation in order that a compiler never has get admission to greater than a small a part ofthis system. In lining a function for which you don't know the supply seems feasible given advanced linker and optimizer technology, but such era wasn't to be had at the time (and still isn't in maximum environments).

2.5 The Linkage Model

The issue of the way one after the other compiled applications are connected together is vitalfor any programming language and, to a degree, determines the capabilities the language can provide. One of the crucial influences at the development of C with instructions and C++ wasthe choice that

1. Separate compilation ought to be possible with traditional C/FORTRAN UNIX/DOS fashionlinkers.
2. Linkage must in precept be kind secure.
3. Linkage must no longer require any form of database (even though one will be used toenhance a given implementation).
4. Linkage to software fragments written in other languages which include C, assembler, andFORTRAN have to be clean and green.

C makes use of "header documents" to ensure constant separate compilation. Declarations of facts structure layouts, capabilities, variables, and constants are located in header files which might be commonly textually blanketed into each supply record that wishes the declarations. Consistency is ensured through putting adequate information in the header files and making sure that the header documents are continuously protected. C++ follows this model up to some extent.

The purpose that format information may be found in a C++ class statement (although it doesn't have to be, is to make sure that the declaration and use of proper local variables is straightforward and efficient. As an instance:

```
Void func( )
{
 Stacks;
 Int c;
 S.push('h');
 C = s.p
```

```
o
p
(
)
;
}
```

Using the stack declaration, even a simple-minded C with
lessons implementation can make certain that no need is
fabricated from free shop for this situation, that the decision of
dad (
) is in lined so that no function name overhead is incurred and
that the non-in lined call of push ( ) can invoke a one by one
compiled characteristic pop (). On this, C++ resembles Ada
[Ichbiah 1979].

The concern for easy-minded implementations became partly a
need due to the lack of assetsfor developing C with classes and
partially a mistrust of languages and mechanisms that required
"smart" strategies. An early components of a layout aim was
that C with lessons "must be implementable without using an
set of rules extra complicated than a linear seek."anyplace that
rule of thumb turned into violated, as inside the case of feature
overloading , it led to semantics that had been more
complicated than each person felt at ease with and usually also
to implementation headaches.
The intention--based totally on my Simula experience--became
to design a language that would be easy sufficient to apprehend
to attract customers and smooth sufficient to enforceto draw
implementers. Only if a fairly simple implementation might be
used by an enormouslynewbie user in a rather unsupportive

programming surroundings to deliver code that as compared favorably with C code in improvement time, correctness, run-time velocity, and code length, ought to C with training, and later C++, anticipate to live to tell the tale in competition with C.

This changed into a part of a philosophy of fostering self-sufficiency among customers. The intention turned into continually and explicitly to develop neighborhood knowledge in all aspects of the use of C++. Most organizations should observe the exact opposite approach. They preserve customers dependent on offerings that generate sales for a critical assist company and/or consultant. In my view, this contrast is a deep cause for some of the variations between C++ and plenty of different languages.

The decision to work in the pretty primitive and nearly universally available framework of the C linking centers triggered the essential trouble that a C++ compiler must continually paintings with handiest partial records approximately a software. An assumption made about a software ought to likely be violated via a program written the next day in a few other language(inclusive of C, FORTRAN, or assembler) and related in probable after this system has commenced executing. This trouble surfaces in many contexts. It's far tough for an implementation to assure

1. That something is unique,
2. That (kind) information is constant,
3. That something is initialized.

In addition, C offers handiest and feeblest guide for the perception of separate name spacesin order that fending off name space pollutants by way of one at a time written program segments becomes a trouble. Over the years, C++ has tried to stand all of these demanding situations without departing from the essential model and generation that gives portability, but within the C with training days we simply relied on the C technique of header documents.Through the popularity of the C linker got here every other "principle" for the development of C++: C++ is just one language in a system and not a complete device. In different phrases, C++ accepts the function of a traditional programming language with a fundamental difference among the language, the running machine, and other vital elements of the programmer's international. This delimits the position of the language in a way this is difficultto do for a language, inclusive of Smalltalk or Lisp, that became conceived as an entire gadget or environment. It makes it critical that a C++ software fragment can name program fragments written in other languages and that a C++ software fragment can itself be referredto as by means of software fragments written in other languages. Being "just a language" additionally allows C++ implementations to gain immediately from equipment written for different languages.

The need for a programming language and the code written in it to be only a cog in a miles larger gadget is of maximum significance to most business customers, yet such co-existence with other languages and systems was apparently not a primary problem to most theoreticians, might-be perfectionists, and academic users. It was the main reason for the success of C++.

2.6 Static Type Checking

To keep the C code away from breaking, it turned into decided to allow the call of an undeclared feature and not carry out type checking on such undeclared features. This becameof path a main hollow in the type machine, and several attempts had been made to decreaseits importance because the foremost occurrence of programming errors before in the end, inC++, the hollow was closed by using creating a call of an undeclared characteristic unlawful. One simple commentary defeated all tries to compromise, and as a result keep a greater diploma of C compatibility: As programmers found out C with training, they misplaced the capacity to find run-time errors as a result of simple type errors. Having come to depend on the sort checking and type conversion furnished by means of C with instructions or C++, theylost the capacity to quickly locate the "silly mistakes" that creep into C packages via the lack of checking. Similarly, they didn't take the precautions towards such silly errors that exact C programmers take as a count of path. After all, "such errors do not take place in C with training." for that reason, as the frequency of run-time mistakes caused by uncaught argument kind mistakes is going down, their seriousness and the time needed to discover them goes up. The result turned into critically annoyed programmers worrying similarly tightening of the kind machine.

The maximum thrilling test with "incomplete static checking" became the method of permitting calls of undeclared features but noting the type of the arguments used in order that a consistency test will be completed when in addition calls have been visible. Whilst Walter brilliant many years later independently discovered this trick, he named it "auto prototyping," the usage of the ANSI C term prototype for a function assertion. The revel in turned into that auto prototyping stuck many mistakes and to begin with multiplied aprogrammer's self-belief within the type machine. However, on the grounds that regular errors and mistakes in a characteristic known as simplest once in a compilation had been now not stuck, auto prototyping ultimately destroyed programmer confidence within the type checker and precipitated a feel of paranoia even worse than that was witnessed in C or BCPLprogrammers.

C with lessons brought the notation f (void) for a characteristic f that takes no arguments as a comparison to f ( ) that in C announces a function that could take any quantity of arguments of any type with none type test. My users soon convinced me, but, that the f (void) notation wasn't very fashionable, and that having functions declared f ( ) be given arguments wasn't very intuitive. Consequently, the end result of the experiment changed into to have f ( ) suggest a characteristic f that takes no arguments, as any beginner might count on. It took support from each Doug mciiroy and Dennis Ritchie for me to build up braveness to make thisdamage from C. Most effective when they used the phrase abomination approximately f (void) did I dare provide f ( ) the apparent which means. However, to this day C's kind regulations are a great deal laxer than C++'s and any use of f ( ) as a characteristic statementbetween the Two languages is incompatible.

2.7 Syntax Problem

In C, the call of a shape, a "shape tag," should continually be preceded with the aid of the key-word struct. For instance

Struct buffer a;   /* 'struct' is vital in C */

In the context of C with classes, this had annoyance for some time as it made user-describedkinds second kind residents syntactically. The call of a struct or a category is now a type call and requires no unique syntactic identity:

Buffer a;   // C++

The ensuing fights over C compatibility lasted for years.

2.8 Derived Classes

The C with classes concept becomes supplied without any shape of run-time support. In particular, the Simula (and C++) concept of a virtual function became missing. The reason for this turned into the purpose, of educating human beings how to use them and, even extra, the persuasion of people that a virtual feature is as green in time and area as an everyday characteristic, as usually used. Often human beings with Simula and Smalltalk revel innonetheless don't quite believe that until they've had the C++ implementation explained to them in element--and plenty of still harbor irrational doubts after that. Even without virtual capabilities, derived lessons in C with classes have been beneficial for constructing new statistics systems out of antique ones and for associating operations with the resulting sorts. In particular, as explained in Stroustrup [1980] and Stroustrup [1982], theyallowed list lessons to be defined, and also task training.

In the absence of digital capabilities, a consumer may want to use gadgets of a derived elegance and treat base instructions as implementation info (only). Alternatively, a specific kind field may be introduced in a base class and used together with express kind casts. The previous approach changed into used for responsibilities where the user handiest sees particular derived venture classes and "the system" sees most effective the undertaking basetraining. The latter approach was used for diverse software instructions wherein, in impact, a base magnificence turned into used to put in force a variant report for a hard and fast of derived training. A great deal of the effort in C with instructions and later C++ has been to make certain that programmers need not write such code.

2.9 Protection Model

Function were made to be able to declare in public parts of the class or by specifying a function or a category as a friend. Initially, simplest instructions may be pals, consequently granting get right of entry to all member features of the friend magnificence, but later it turned into observed handy so that it will grant get right of entry to (friendship) to individualfeatures. Specifically, it changed into found useful if you want to furnish get entry to global functions. A friendship declaration turned into seen as a mechanism much like that of one safety domain granting a study-write capability to some other.

2.10 Run-Time Guarantees

The access control mechanisms defined above honestly prevent unauthorized get right of entry to. A second form of guarantee supplied by "special member features," which includesconstructors, that have been identified and implicitly invoked through the compiler. The idea was to permit the programmer to establish guarantees, once in a while called "invariants," that different member functions may want to rely upon. Curiously sufficient, the preliminary implementation contained a feature that isn't always provided via C++ but is regularly asked. In C with lessons, it became feasible to define a characteristic that might implicitly be known as earlier than each name of every member function (except the constructor) and another that might be implicitly referred to as earlier than every return from every member function. They were called call and return functions. They were used to provide synchronization for the monitor class inside the original task library [Stroustrup 1980b]:

Class monitor: object
{
/* ... */
C
a
l
l
(
)
(
/
*
p
i
c
k
l
o
c
k
*

/
)
R
e
t
u
r
n
(
)
(
.
/
*
r
e
l
e
a
s
e
l
o
c
k
*
/
)
}:

These are comparable in cause to the CLOS: before and: after methods. Name and return capabilities were eliminated from the language due to the fact no person used them and because no one persuaded human beings that call and return functions had essential uses. In 1987, Mike Tiemann cautioned an alternative solution known as "wrappers" [Tiemann 1987].

Chapter 3

3.1 From C with Classes to C++

Throughout 1982, it became clear that C with classes became a "medium success" and might remain so till it died. So, a medium was defined for fulfillment as something so beneficial that it without problems paid for itself and its developer, but not so attractive and beneficial that it'd pay for a help and improve organization. Accordingly, persevering with C with classes and its C preprocessor implementation would condemn to aid C with classes' use indefinitely. So, there were only two methods derived out of this dilemma:

1. Stop supporting C with classes, in order that the customers could need to pass elsewhere.

2. Develop a new and better language based on my experience with C with classes that would serve a huge sufficient set of users to pay for aid and development enterprise hence at that time it was observed that 5000 commercial users are essential minimum.

The success of C with classes became an easy outcome of assembly its layout purpose: C with instructions did help arrange a large elegance of applications considerably higher than C, without the lack of run-time performance and without requiring enough cultural adjustments to make its use unfeasible in groups that had been unwilling to undergo foremost changes. The elements restricting its success have been partially the limited set of latest facilities presented over C, and partially the preprocessor generation used to implement C with lessons. There simply wasn't sufficient assist in C with classes for individuals who have been willing to invest tremendous efforts to acquire matching advantages: C with classes become a vital step in the proper route, but simplest one small step.

The resulting language was at the start nonetheless referred to as C with lessons, however after a polite request from control it changed into given the name C84. The motive for the naming becomes that people had taken to calling C with classes "new C," after which C. This closing abbreviation caused C being known as "simple C," "immediately C," and "vintage C." The call C84 turned into used only for a few months, partially as it changed into unsightly andinstitutional, in part because there would nonetheless be confusion if humans dropped the "84." ideas for a brand-new call were asked and picked C++ because it became brief, had pleasant interpretations, and wasn't of the form "adjective C." In C, ++ can, depending on context, be examine as "next," "successor," or "increment," even though it's far usually suggested "plus". The call C++ and its runner up ++C are fertile assets for jokes and puns-- almost all of which had been regarded and favored before the name become chosen. The callC++ turned into cautioned by Rick Mascitti. It changed into first used in Stroustrup [1984b] wherein it become edited into the final copy in December 1983.

3.2 Cfront

The Cfront compiler front-end for the C84 language changed into designed and carried out via me between the spring of 1982 and the summer time of 983. The primary person outside the computer science research center, Jim Coplien, obtained his replica in July of 1983. Jim changed into in a group that were doing experimental switching paintings with C with Classes in Bell Labs in Naperville, Illinois, for some time.

In that equal time period designed C84, drafted the reference guide published January 1, 1984 [Stroustrup 1984a], designed the complicated variety library and implemented it, collectively with Eeonie Rose[Rose]984], designed and implemented the first string class together with Jonathan Shopiro, maintained and ported the C with classes implementation, and supported the C with classes customers and helped them grow to be C84 users. Cfront turned into (and is) a traditional compiler the front-end, performing a whole take a look at of the syntax and semantics of the language, building an inner representation of its input, reading and rearranging that illustration, and subsequently producing output suitable for a few code generators. The internal illustration changed into (is) a graph with one symbol table in keeping with scope. The overall strategy is to study a supply document one worldwide statement at atime and bring output handiest when a whole global declaration has been completelyanalyzed.

The agency of Cfront is reasonably traditional, besides perhaps for the use of many symbol tables instead of just one. Cfront become at the beginning written in C with Classes and soon transcribed into C84 so that the very first running C++ compiler turned into executed in C++.Even the first model of Cfront used classes heavily, however no virtual capabilities because they were not available at the project start.

The maximum uncommon for its time issue of Cfront become that it generated C code. This has brought on no cease of confusion. Cfront generated C due to the fact. Ought to easily have generated some inner back-give up format or assembler from Cfront, however that turned into not what customers wanted. In reaction to this need, concluded that using C as a common input format to a massive variety of code turbines became the only reasonable preference. The strategy of constructing a compiler as a C generator has later become prettypopular, in order that languages consisting of Ada, CLOS, Eiffel, Modula-three, and Smalltalk have been applied that way. C compiler is used as a code generator most effective. Any mistakes message from the C compiler reflects a mistake inside the C compiler or in Cfront, however no longer within the C++ source textual content. Every syntactic and semantic blunder is in precept caught by usingCfront, the C++ compiler front-give up. There has been a protracted history of bewildermentabout what Cfront changed into/is. It's been known as a preprocessor because it generates C,and for humans within the C network (and some other place) that has been taken as evidencethat Cfront changed into an alternatively easy program something like a macro preprocessor.Humans have accordingly "deduced" (wrongly) that a line-for-line translation from C++ to C is possible, that symbolic debugging at the C++ level is impossible when Cfront is used, that code generated by using Cfront should be not so good as code generated with the aid of "actual compilers," that C++ wasn't a "actual language," and so on.

Cfront is just a compiler front-end and can by no means be used for real programming itself. It uses a driver force to run the source document through the C preprocessor, Cpp, then run the output of Cpp through Cfront, and the output from Cfront through a C compiler:

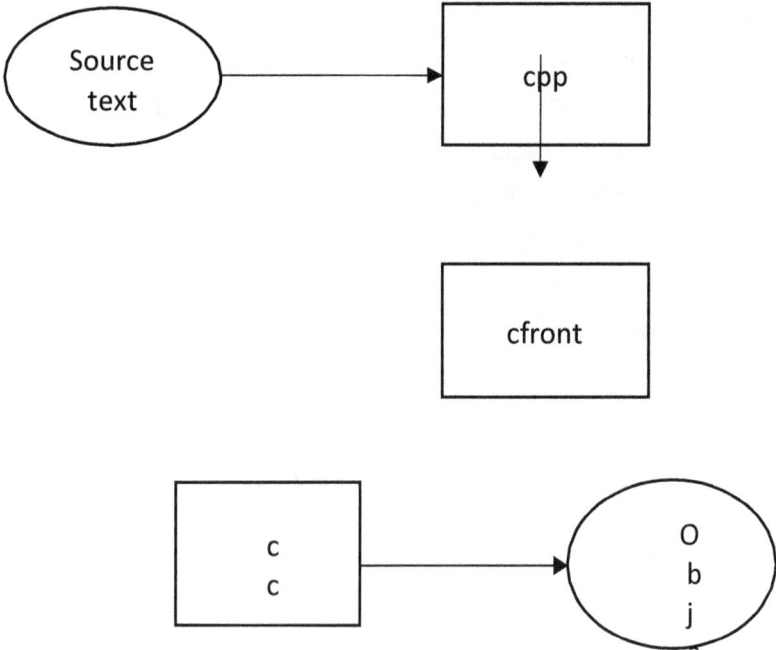

```
  ┌──────────┐              ┌──────────┐
 (  Source   )────────────▶ │   cpp    │
 (   text    )              │     │    │
  └──────────┘              └─────┼────┘
                                  ▼

                            ┌──────────┐
                            │  cfront  │
                            └──────────┘

  ┌──────────┐              ┌──────────┐
  │    c     │              (    O     )
  │    c     │────────────▶ (    b     )
  └──────────┘              (    j     )
                            └──────────┘
```

Similarly, the driver ought to make certain that dynamic (run-time) initialization is carried out. In Cfront 3.0, the driver became but extra intricate as automated template instantiation.

As stated, it was decided to stay within the constraints of traditional linkers. But there was one constraint that was felt too tough to live with: most traditional linkers had a very low restriction on the variety of characters that may be used in external names. A limit of 8 characters changed into not unusual, and 6 characters and one case most effective are guaranteed to work as outside names in Classical C; ANSI/ISO C accepts that restrict also. For the reason that the call of a member feature includes the name of its class and that the type of an overloaded characteristic needs to be meditated inside the linkage manner in some way or other. Cfront uses encodings to enforce type safe linkage in a way that makes a restriction of 32 characters too low for comfort and even 256 is a piece tight at instances. Within the meantime, systems of hash coding of long identifiers were used with archaic linkers, but that turned into by no means completely great.

Versions of C++ are often named by means of Cfront launch numbers. Launch 1.0 was the language as described in "The C++ Programming Language".

Releases 1.1 (June 1986) and 1.2 (February 1987) had been typically bug restoration releaseshowever also introduced hints to individuals and guarded contributors. Release 2.0 become a primary clean-up that still brought multiple inheritance in June 1989. Release 2.1 (April 1990) changed into more often than not a computer virus restore release that introduced Cfront (nearly) into line with the definition within the ARM. Release 3.0 (September 1991) delivered templates as detailed within the ARM. Launch 4.0 is predicted to feature exceptiondealing with as distinctive within the ARM.

3.3 Language Feature Details

1. References
2. Function name and operator overloading
3. Virtual functions
4. Improved type checking
5. User-controlled free-store memory control
6. Constants (const)

3.4 Virtual Functions

The most obvious new characteristic in C++, and clearly the one that had the finest impact onthe fashion of programming one should use for the language, was virtual functions. The idea become borrowed from Simula and presented in a shape that intended to make easy and efficient implementation.

The motive for virtual functions was presented in Stroustrup [1986b] and [1986c]. To emphasize the important position of virtual capabilities in C++ programming, i can be quote:"An abstract data type defines a form of black box. As soon as it's been described, it does nownot truly interact with the rest of the program. There's no way of adapting it to new uses except via modifying its definition. This could cause severe inflexibility. Keep in mind defininga type form for use in a portrait's device. Count on for the instant that the system has to aid circles, triangles, and squares. Expect additionally that you have a few classes:

C
l
a
s
s
p
o
i
n
t

/
*
.
.
.
*
/
)
;
C
l
a
s
s
c
o
l
o
r
/
*
.
.
.
*
/
;
You might
outline a
shape like
this: Enum

```
kind (
circle,
triangle,
rectangula
r };Class
shape (
point
center;
Color col;
```

Kind k;
/
/
r
e
p
r
e
s
e
n
t
a
t
i
o
n
o
f
s
h
a
p
e
P
u
b
l
i
c
:

```
Point where () ( return center; }
Void move (point to)  {center = to; draw();
)
Void draw();
Void rotate(int);
// extra operations
```

};

The "type field" k is important to permit operations which include draw ( ) and rotate
Of shape they are managing (in a Pascal-like language, one would possibly use a variant recordwith tag k). The function draw ( ) is probably described like this:

V
o
i
d
s
h
a
p
e
:
:
d
r
a
w
(
)
{
S
w
i
t
c

h
(
k
)
{
c
a
s
e
c
i
r
c
l
e
:
/
/
d
r
a
w
a
c
i
r
c
l
e
B
r

e
a
k
;
c
a
s
e
t
r
i
a
n
g
l
e
:
/
/
d
r
a
w
a
t
r
i
a
n
g
l

```
e
b
r
e
a
k
;
C
a
s
e
s
q
u
a
r
e
:
// draw a square
}}
```

This is a large number. Capabilities along with draw ( ) have to "know about" all the sorts of shapes there are. Consequently, the code for this type of characteristic grows every time a brand-new shape is brought to the device. If you define a brand-new shape, every operationon a shape have to be tested and (probably) modified. You aren't able to upload a new shape to a device except you have got get entry to the supply code for every operation. Because adding a brand-new form entails "touching" the code of every essential operation on shapes,it requires awesome skill set and potentially introduces bugs into the code dealing with other (older) shapes. The choice of illustration of precise shapes can get significantly cramped through the requirement that (as a minimum a number of) their representation should suit into the normally constant sized framework presented through the definition of the overall kind shape.

The Simula inheritance mechanism presents a solution. First, specify a class that defines the general properties of all shapes:

C
l
a
s
s
s
h
a
p
e
{
p
o
i
n
t
c
e
n
t
e
r

;
C
o
l
o
r
c
o
l
;
// ... Public :

```
Point where() { return center; }
Void move(point to) { center =
to; draw(); ) virtual void draw();
Virtual void rotate (int) ;
// ...
};
```

The functions for which the calling interface may be described, however where the implementation cannot be described besides for a selected shape, had been marked "digital"(the Simula and C++ time period for "can be redefined later in a category derived from this one"). Given this definition, standard capabilities manipulating shapes were written:

```
Void rotate_all(shape** v, int size, int angle)
// rotate all members of vector "v"
of size "size" "angle" degrees (For
(int i = O; i < size; i++) v[i]
.rotate(angle);
)
```

To outline a particular shape, we have to say that it is a shape and specify its particular properties (which includes the virtual function):

C
l
a
s
s
c

i
r
c
l
e
:
p
u
b
l
i
c
s
h
a
p
e
(
l
n
t
r
a
d
i
u
s
;
p
u
b

l

i

c

:

Void draw ( ) { /* ... */ };
Void rotate(int) {} // yes, the null function};

In C++, c1 as circle is said to be derived from class shape, and class shape is said to be a baseof class circle. An alternative terminology calls circle and shape subclass and superclass, respectively.

The key implementation concept became that the set of virtual functions in a category definesan array of pointers to functions, in order that a name of a digital characteristic is virtually an oblique characteristic name thru that array. There is one array in step per class and one pointer to such an array in every object of a class that has virtual functions.

Designed software wouldn't want the extensibility and openness provided by using virtual functions, in order that right evaluation might show which non-virtual functions will be referred to as at once. Therefore, the argument went, virtual functions have been actually a shape of inefficiency. But virtual functions were added in the language anyway.

3.5 Overloading

Several human beings had requested for the capacity to overload operators. Reluctant pointsto not to add overloading in C++:

1. Overloading become reputed to be tough to enforce so that compilers would grow totremendous size.
2. Overloading was reputed to be tough to educate and hard to outline precisely in orderthat manuals and tutorials would develop to giant length.
3. Code written the use of operator overloading became reputed to be inherently inefficient.
4. Overloading changed into reputed to make code incomprehensible.

If all of these conjectures were false, then overloading might remedy some real problems forC++ users. There had been folks that would really like to have complex numbers, matrices, and APL-like vectors in C++. There were folks that would really like range-checked arrays, multi-dimensional arrays, and strings in C++. There have been at the least two separate programs for which people desired to overload logical operators which includes I(or), & (and),and ^ (distinctive or). The way I saw it, the listing turned into long and might grow with the scale and the variety of the C++ consumer populace. My answer to [4], "overloading makes code difficult to understand," become that several of the programmers, whose opinions werevalued and whose experience became measured in many years, claimed that their code would become cleaner in the event that they had overloading. So, what if you'll write difficult to understand code with overloading? It's miles possible to put in writing obscure code in any language. Its topics extra how a feature may be used properly than how it could be misused.First it was discovered that use of class with over-loaded operators, such as complex and string, changed into pretty easy and didn't placed a main burden at the programmer. Subsequent a guide released, the guide sections to prove that the delivered complexity wasn'ta serious problem; the forty-two-page manual needed less than a page and a half more. So, the first implementation in hours using most effective 18 strains of extra code in Cfront.

Certainly, these kinds of problems were not genuinely tackled in this strict sequential order. But, the focal point of the paintings did begin with application issues and slowly drifted to implementation troubles.

In retrospect, the complexity of the definition and implementation problems and compounded these troubles by seeking to isolate overloading mechanisms from the relaxation of the language semantics. The latter become done out of erroneous worry of perplexing customers.

Overload print ( ) ;

Should precede declarations of an overloaded

function print, which includes Void print(int) ;

Void print (const char*) ;

Additionally, insisted that ambiguity manipulate should take place in two stages so that resolutions concerning built-in operators and conversions would constantly take precedence over resolutions related to person-defined operations. Maybe the latter changed into inevitable, given the priority for C compatibility and the chaotic nature of the C conversion regulations for built-in types. These conversions do now not constitute a lattice; for instance, implicit conversions are allowed each from into to drift and from waft to int. But, the policiesfor ambiguity resolution have been too complicated, triggered surprises, and needed to be revised for launch 2.0.

Requiring express overload declarations become undeniable wrong and the requirement become dropped in release 2.0.

3.6 References

References had been delivered basically to aid operator overloading. C passes each function argument by means of value, and in which passing an object by using value might beinefficient or beside the point the user can bypass a pointer. This strategy doesn't work whereoperator overloading is used. If so, notational comfort is vital so that a user can't anticipate to insert address of operators if the objects are massive.

Troubles with debugging ALGOL 68 convinced that having references that did not trade whatobject they stated after initialization, was an excellent component. Due to the fact C++ has each recommendations and references, it does not want operations for distinguishing operations at the reference itself from operations on the object cited (like Simula), or the sortof deductive mechanism hired by using ALGOL 68.

It's essential that const references can be initialized by way of non-lvalues and lvalues of types that require conversion. Specifically, that is what lets in a FORTRAN feature to be referred toas with a steady:

Extern "Fortran" go with the " float sqrt (const flow&); // °&' means reference sqrt(2); // callby way of reference

Jonathan Shopiro was deeply worried inside the discussions

that brought about the introduction of references. Similarly, to the obvious makes use of references, together with argument, taken into consideration the ability to apply references as return sorts critical. This allowed to have a very simple index operator for a string class:

```
Class String { // ...
Char& operator [] (int index); // subscript operator // return a reference
};
Void f(String& s)
{
Char cl = ...
S[i] = cl; //
assign to
operator [] 's
result // ...
Char c2 = s[i];
// assign
operator [] 's
result
}
```

The consideration permitting separate features for left-hand and right-hand side use of afeature but considered the use of references the easier opportunity, even though this implied to introduce extra "helper classes" to resolve a few issues where returning a simple referencewasn't sufficient.

3.7 Constants (const)

In operating systems, it's far common to have get access to a few pieces of memory controlled without delay or circuitously through bits: one which indicates whether or not a user can writeto it and one that shows whether a user can read it. This concept appeared without delay relevant to C++ and [taken into consideration allowing every kind to be distinctive read onlyor write only. The notion is focused on specifying interfaces instead of on offering symbolic constants for C. Clearly, a read-only value is a symbolic constant, but the scope of the conceptis far greater. To start with, it was proposed to read-only but now not read-only pointers. Sometime later, the ANSI C committee (X3J]!) Was formed and the const proposal resurfacedthere and have become a part of ANSI/[SO C.

But, within the interim [had experimented similarly with const in C with lessons anddiscovered that const become a beneficial opportunity to macros for representing constants best if an international consts were implicitly local to their compilation unit. Best if so, shouldthe compiler without problems deduce that their value truly did not alternate and allow simple consts in steady reviews and for that reason avoid allocating space for such constantsand use them in constant expressions. C did no longer undertake this rule. This makes constsfar less useful in C than in C++ and leaves C depending on the preprocessor in which C++ programmers can use nicely typed and scoped consts.

3.8 Memory Management

Long earlier the primary C with classes program was written, Bjarne Stroustrup knew that free store (dynamic memory) would be used extra closely in a language with classes than in traditional C programs. This turned into the reason for the introduction of the new and deleteoperators in C with Classes. The new operator that each allocates memory from the free storeand invokes a constructor to ensure initialization turned into borrowed from Simula. The delete operator was a necessary complement because Bjarne Stroustrup did not want C withlessons to rely on a garbage collector. The argument for the new operator may be summarizedlike this. Could you rather write:

```
X* p = new X(2);
Struct X * p =
(struct X *)
malloc(sizeof(struc
t X));If (p == O)
error("memory
exhausted");
P->init (2) ;
```

And in which version are you maximum likely to make a screw up? The arguments towards it,which had been voiced pretty much load at the time, have been "however we don't actually need it," and "but a person can have used new as an identifier." both observations are correct,of course.

Introducing "operator new" for this reason made the use of free save extra convenient and much less errors prone. This expanded its use even in addition so that the C free store allocation habitual m a l l o c ( ) used to enforce new became the most not unusual overall performance bottleneck in actual systems. This turned into no real wonder both; the only hassle become what to do approximately it. Having real packages spend 50 percentage or more in their time in malloc ( ) wasn't desirable.

Bjarne Stroustrup discovered consistent with-class allocators and deallocators very powerful.The fundamental idea is that free store memory usage is dominated by way of the allocation and deallocation of plenty of small gadgets from only a few instructions. Take over the allocation of those items in a separate allocator and you can shop both time and space for those objects and also lessen the quantity of fragmentation of the overall free store. The mechanism provided for 1.0, "assignments to this," changed into too low level and error prone and turned into changed with the aid of a cleaner solution in 2.0.

Observe that static and automatic (stack allotted) objects had been continually feasible and that the only memory control strategies relied closely on such objects. The string class became a standard example; here string objects are normally on the stack in order that they require no explicit memory management, and the free store they depend on is controlled completely and invisibly to the user with the aid of the String member functions.

3.9 Type Checking

The C++ kind checking rules were the result of experiments with the C with classes. All characteristic calls are checked at bring together time. The checking of trailing arguments can be suppressed through explicit specification in a feature statement. That is crucial to allow C'sprint f ():

Int printf(const char* ...) ;
// accept any argument after
// the initial character string
// ...
Printf("date: %s %d 19%d\n", month, day, year); / / maybe right

Numerous mechanisms have been supplied to relieve the withdrawal signs and symptoms that many C programmers feel when they first revel in strict checking. Overriding kind checking using the ellipsis changed into the most drastic and least advocated of these. Characteristic name overloading and default arguments [Stroustrup 1986b] made it possible to provide the arrival of a single function taking a selection of argument lists withoutcompromising type safety. The stream I/O system demonstrates that the weak checking wasn't essential even for I/O.

Chapter 4

4.1 C++ 2.0

Now (mid 1986), the direction for C++ become set for all who cared to see. The important thing design decisions were made. The course of the destiny evolution become for parameterized types, multiple inheritance, and exception handling. Tons experimentation and adjustment based totally on experience was needed, however the glory days had been over. C++ had by no means been stupid putty, however, there has been now no realopportunity for radical change. For suitable and horrific, what become carried out become accomplished. What become left turned into a super quantity of strong work. At this point C++ had about 2,000 users international.

This was the factor in which the plan as at first conceived by way of Steve Johnson and BjarneStroustrup was for a development and support organization to take over the everyday work at the tools (in general Cfront), for that reason releasing Bjarne Stroustrup to work on the new functions and the libraries that had been expected to depend upon them. This become additionally the factor in which Bjarne Stroustrup predicted first AT&T, and then others, couldbegin to construct compilers and different tools to ultimately make Cfront redundant.

Virtually, they had already started, but the precise plan turned into soon derailed due to management indecisiveness, ineptness, and absence of focus. A project to develop a trendy C++ compiler diverted interest and resources from Cfront protection and improvement. A plan to ship a release 1.3 in early 1988 absolutely fell thru the cracks. The net impact becomethat they had to wait until June of 1989 for launch 2.0, and that even though 2.0 become significantly better than release 1.2 in almost all ways, 2.0 did not provide the language capabilities mentioned inside the "whatis paper," and therefore a substantially progressed and prolonged library wasn't part of it.

Some of the folks who prompted C with classes and the original C++ endured to assist with the evolution in diverse methods. Phil Brown, Tom Cargill, Jim Coplien, Steve Dewhurst, Keith Gorlen, Laura Eaves, Bob Kelley, Brian Kernighan, Andy Koenig, Archie Lachner, Stan Lippman,Larry Mayka, Doug McIlroy, Pat Philip, Dave Prosser, Peggy Quinn, Roger Scott, Jerry Schwarz, Jonathan Shopiro, and Kathy Stark were explicitly recounted in [Stroustrup 1989b].

Stability of the language definition and its implementation

become taken into consideration important. The functions of 2.0 have been fairly simple modifications of the language primarily based on experience in with the 1. * releases. The maximum important element oflaunch 2.0 changed into that it multiplied the generality of the individual language features and advanced their integration into the language.

4.2 Feature Overview

1. Better resolution of overloaded functions.

2. Recursive definition of assignment and initialization
3. Multiple inheritance
4. Type-safe linkage
5. Abstract classes
6. Better facilities for user-defined memory management
7. P r o t e c t e d members (first provided in release 1.2)
8. Pointers to members (first provided in release 1.2)
9. Static member functions,
10. Overloading of operator ->, and
11. Const member functions

Maximum of those extensions and refinements represented experience won with C++ and could not have been added earlier without greater foresight than Bjarne Stroustrup possessed. Evidently, integrating these functions involved full work, but it become very unlucky that this became allowed to take precedence over the completion of the language as outlined in the "whatis" paper.

Most capabilities enhanced the protection of the language in some way or different. Cfront

2.0 checked the consistency of characteristic sorts throughout separate compilation units (type safe linkage), made the overload resolution guidelines order independent, and also ensured that greater calls have been taken into consideration ambiguous. The notion of constwas made extra comprehensive, and pointers to members closed a loophole within the type system and provided explicit class-specific memory allocation and deallocation operations to make the error-susceptible "assignment to this" technique redundant.

To a few people, the most critical "feature" of release 2.0 wasn't a feature in any sense but asimple space optimization.

From the start, the code generated with the aid of Cfront tended to be pretty suitable. As late as 1992, Cfront generated the fastest strolling code in a benchmark used to assess C++ compilers on a Sparc. There were no tremendous improvements in Cfront's code era because launch 1.0. However, launch 1. * changed into wasteful because every compilation unit generated its personal set of digital feature tables for all the lessons utilized in that unit. This could result in megabytes of waste. At the time (approximately 1984), Bjarne Stroustrup considered the waste essential within the absence of linker guide and asked for such linker assist. By means of 1987 that linker support hadn't materialized. Therefore, Bjarne Stroustrup re-thought the problem and solved it by the simple heuristic of laying down the virtual characteristic table of a category right next to its first non-digital non-inline function.

4.3 Multiple Inheritance

In most of the people's minds, a couple of inheritance, the ability to have or more direct baseclasses, is the function of 2.0. Bjarne Stroustrup disagreed on the time because Bjarne Stroustrup felt that the sum of the upgrades to the sort device have been of far more sensible importance. Also, adding a couple of inheritance in 2.0 changed into a mistake. More than one inheritance belongs in C++ however is a way less critical than parameterized sorts. Because it passed off, parameterized types inside the form of templates handiest regarded inrelease 3.0. There had been more than one reasons for deciding on to work on multiple inheritance at the time: The design became further superior and the implementation could be achieved within Cfront. Every other thing turned into simply irrational. Nobody doubted that Bjarne Stroustrup should put in force templates successfully. Hence multiple inheritance,however, become extensively speculated to be very hard to implement correctly. For that reason, multiple inheritance seemed extra of an undertaking, and due to the fact Bjarne Stroustrup had taken into consideration it as early as 1982 and located an easy and efficient implementation method in 1984, Bjarne Stroustrup could not face up to the mission. I suspectthat that is the handiest case in which style affected the series of events.

In September of 1984, Bjarne Stroustrup supplied the C++ operator overloading mechanism at the IFIP WG2.four conference in Canterbury [Stroustrup 1984c]. There Bjarne Stroustrup met Stein Krogdahl from the university of Oslo who became just completing an offer for adding multiple inheritance to Simula [Krogdahl 1984]. His thoughts became the basis for theimplementation of regular a couple of base lessons in C++. He and that Bjarne Stroustrup laterdiscovered out that the thought become almost identical to an idea for providing multiple inheritance in Simula that were taken into consideration by way of Ole-Johan Dahl in 1966 and rejected because it might have complicated the Simula rubbish collector [Dahl 1988].

The unique and essential purpose for thinking about a couple of inheritance changed into virtually to permit training to be blended into one in this kind of way that items of the resulting class would behave as items of either base class [Stroustrup 1986c]:

A fairly fashionable example of the usage of multiple inheritance might be to provide library classes displayed and task for representing items under the control of a show manager and co-routine under the control of a scheduler, respectively. A programmer ought to then createlessons such as

Class my_displayed task : public displayed, public task { // ...
};
Class my_task : public task { // not displayed // ...
};
Class my_displayed : public displayed { // not a task // ...
};

The usage of (simplest) single inheritance handiest of those three alternatives would be opento the programmer.

The implementation requires little greater than remembering the relative offsets of the Cask and displayed objects in a my_displayed_task object. All the gory implementation info had been defined in Stroustrup [1987a]. Similarly, the language layout ought to specify how ambiguities are treated and what to do if a class is special as a base magnificence extra thanas soon as in a derived class:

Ambiguities am handled ~ compile time:
Class A { public: void f(); /* ... */ };
Class B { public: void f(); /* ... */ };
Class C
:public A,
public B { /*
no f() ... */ };
Void g() { C*
p;
P->f(); // error: ambiguous }

In this, C++ differs from the item-oriented Lisp dialects that support multiple inheritance. Basically, Bjarne Stroustrup rejected all types of dynamic resolution past using virtual featuresas incorrect for a statically typed language under extreme efficiency constraints. Maybe,Bjarne Stroustrup must at this factor have revived the notion of call 1 and return functionsto mimic the CLOS: before and : after methods. But human beings were already traumaticabout the complexity of the multiple inheritance mechanisms and i am continually reluctantto re-open old wounds.
A couple of inheritance in C++ became debatable [Cargill 1991; Carroll 1991; Waldo 1991; Sakkinen 1992] for several motives. The arguments against it targeted across the real and imaginary complexity of the concept, the utility of the idea, and the effect of more than one inheritance on other extensions and tool building. Further, proponents of more than one inheritance can, and do, argue over exactly what more than one inheritance

is meant to be and the way its miles quality supported in a language. I assume, as I did then, that the essentialflaw in these arguments is that they take more than one inheritance far too significantly. Multiple inheritance would not resolve all your troubles; however, it doesn't need to becauseit is pretty cheap, and from time to time it is very convenient to have. Grady sales space [Booch 1991] expresses a slightly more potent sentiment: "more than one inheritance is likea parachute, you do not want it very often, however while you do it's essential."

4.4 Abstract Classes

The final feature delivered to 2.0 earlier than it shipped turned into virtual. Past due change to releases are never famous and overdue changes to the definition of what will be shipped are even less so. Bjarne Stroustrup understood that several individuals of control concept I had lost touch with the real global when I insisted on this selection.

A commonplace criticism about C++ turned into (and is) that private data is seen and that after private statistic is modified then code using that classes have to be recompiled. Regularly this complaint is expressed as "abstract types in C++ aren't really abstract." What Bjarne Stroustrup hadn't found out was that many human beings idea that because they could put the illustration of an object within the private section of a class announcement then they virtually had to placed it there. This is actually incorrect (and that there was hassle for years).

If you do not need a representation in a class, accordingly, making the magnificence an interface only, you then really postpone the specification of the representation to some derived magnificence and outline simplest digital functions. For example, one can define a setof T pointers like this:

```
Class set { public :
Virtual        void
insert(T*); virtual
void remove(T*) ;
Virtual         int
is_member(T*);
V
i
r
t
u
a
l
T
*
f
i
r
s
t
(
)
;
V
i
```

```
r
t
u
a
l
T
*
n
e
x
t
(
)
;
V
i
r
t
u
a
l
-
s
e
t
(
)
{
}
}
;
```

This provides all the information that people need to use a set, except that whoever actuallymade a set must know something about how a few particular kind of set is represented
For example, given

```
Class slist_set : public set,
private slist { slink*
current_elem;Public:
Void insert(T*);
Void
remove(
T*); int is
member
(T*);
Virtual
T* first();
virtual
T*
next();
Slist_set(
) : slist(),
current_
elem(0) {
});
```

We can create slist_set gadgets that may be used as sets through users who've in no way heard of a
Slist_set. The handiest problem become that in C++, as defined earlier than 2.0, there has been no express manner of pronouncing: "The set class is just an interface: its functions neednot be defined, it is an error to create objects of class set,

and each person who derives a category from set must outline the virtual features specified in set." release 2.0 allowed a category to be declared explicitly abstract via declaring one or more of its virtual capabilities"natural" using the syntax = 0:

Class set { // abstract class public :
Virtual void insert(T*) = 0; // pure virtual function virtual void remove(T*) = 0;
// ..o);

The =0 syntax is not exactly wonderful; however, it expresses the preferred perception of a natural virtual characteristic in a way that is terse and fits the use of 0 to mean "not anything"or "no longer there" in C and C++. The opportunity, introducing a new keyword, say natural, wasn't a choice. Given the opposition to abstract classes as a "late and unimportant exchange," Bjarne Stroustrup had in no way simultaneously have triumph over the traditional, strong, full-size, and emotional competition to new key phrases in components of the C and C++ community.

The significance of the abstract class idea is that it allows a cleaner separation among a user and an implementor than is viable without it. This limits the amount of recompilation vital after an alternate and also the quantity of records important to compile a median piece of code. Via reducing the coupling between a consumer and an implementor, abstraction classes provided a solution to user complaining about lengthy compile duration, and also serve libraryproviders who need to worry approximately the effect on users of adjustments to a library implementation.

4.5 Libraries

The first actual code written in C with instructions was the task library [Stroustrup 1980b], which supplied Simula like concurrency for simulation. The first real packages have been simulations of community site visitors, circuit board layout, and so forth, the use of the project library. The project library continues to be heavily used today. The usual C library changed into available from C++, without overhead or worry in comparison with C, from day one. So are all different C libraries. Classical data types, along with character strings, range checked arrays, dynamic arrays, and lists, have been a few of the examples used to design C++ and check its early implementations.

The early work with field classes such as list and array were significantly hampered through the shortage of assist for a manner of expressing parameterized types in C with classes and inC++ up until version 3.0. In the absence of proper language assist later supplied within the shape of templates, we needed to make do with macros. The first-rate that can be said for the C preprocessor's macro centers is that they allowed us to advantage enjoy with parameterized types and support individual and small organization use.

Much of the work on designing classes become finished in cooperation with JonathanShopiro, who in 1983 produced list and string training that noticed extensive use inside AT&Tand are the idea for the instructions currently determined inside the "general additives" library that turned into advanced in ring labs, and is now offered by using USL. The design ofthose early libraries interacted at once with the design of the language and especially with the design of the overloading mechanisms.

4.6 Compilers

The Santa Fe conference marked the announcement of the second one wave of C++ implementations. Steve Dewhurst defined the architecture of a compiler he and others had been building in AT&T's Summit facility, Mike Ball presented some thoughts for what have become the taumetric C++ compiler (greater regularly called the Oregon software program C++ compiler), and Mike Tiemann gave a maximum animated and exciting presentation of ways the GNU C++ he become building might do just about the whole thing and put all different C++ compiler writers out of commercial enterprise. The brand-new AT&T C++compiler by no means materialized; GNU C++ model 1.13 became first launched in Decemberof 1987; and taumetric C++ turned into first shipped in January of 1988.

Till June of 1988, all C++ compiler on computers were Cfront ports. Then Zortech started transport their compiler. The advent of Walter vivid's compiler made C++ "actual" for plenty computer-orientated people for the primary time. Greater conservative human beings reserved their judgment until the Borland C++ compiler in can also of 1990, or maybe Microsoft's C++ compiler in March 1992. DEC launched their first independently advanced C++ compiler in February of 1992 and IBM released their first independently advanced C++ compiler in May of 1992. In all, there are actually extra than a dozen independently evolved C++ compilers.

Similarly, to these compilers, Cfront ports regarded to be everywhere. Specially, sun, HP, Centerline, parcplace, Glockenspiel, and Comeau Computing ship Cfront-based totally products on just about any platform.

4.7 Tools and Environments

C++ was designed to be a possible language in a tool-poor environment. This was partly a need due to the almost complete lack of sources within the early years and the relative poverty in a while. It changed into additionally an aware decision to permit easy implementations and, mainly, easy porting of implementations.

C++ programming environments at the moment are emerging that are a suit for the environments habitually supplied with different object-oriented languages. For example, object works for C++ from parcplace is essentially the great Smalltalk program improvement surroundings tailored for C++, and Centerline C++ (formerly Saber C++) is an interpreter-based totally C++ surroundings, inspired by means of the interlisp environment. This gives C++ programmers the option of using the whizzier, extra expensive, and frequently greater productive environments that have previously handiest been available for different languagesand/or as research toys.

An environment is a framework in which tools can cooperate. There may be now a number of such environments for C++: most C++ implementations on pcs are compilers embedded ina framework of editors, equipment, document systems, standard libraries, and so on. Macappand the Mac MPW is the Apple Mac model of that and ET++ is a public area model within thestyle of the macapp. Eucid's Energize and HP's Softbench are yet different examples.

4.8 Minor Features

The ARM supplied some minor functions that had been not applied until 2.1 releases from AT&T and different C++ compiler carriers. The most obvious of these have been nested classes. Bjarne Stroustrup strongly recommended to revert to the original definition of nested class scopes by comments from external reviewers of the reference guide. Bjarne Stroustrup additionally despaired of ever getting the scope policies of C++ coherent even as the C rule turned into in vicinity.

The ARM allowed users to overload prefix and postfix increment (++) independently. The principle impetus for that got here from those who desired "smart pointers" that behaved exactly like normal recommendations except for some brought work performed "behind the scenes."

4.9 Templates

Within the original design of C++, parameterized types (templates) have been taken into consideration but postponed because there wasn't time to do a thorough activity of exploring the design and implementation troubles. Bjarne Stroustrup first presented templates at the 1988 USENIX C++ conference in Denver:

For plenty of people, the biggest alone trouble using C++ is the shortage of an extensive standard library. A chief hassle in producing any such library is that C++ does now not offer a sufficiently popular facility for defining "container training" together with lists, vectors, and associative arrays. [Stroustrup 1988b]

There are two approaches for supplying such classes/types: one can rely on dynamic typing and inheritance, as Smalltalk does, or one can rely upon static typing and a facility for arguments of kind type. The previous could be very flexible, however carries a high run-time cost, and more importantly, defies tries to use static type checking to seize interface errors.

Consequently, the latter technique turned into chosen.

A C++ parameterized kind is known as a category template. A class template specifies how character training can be constructed much like the way a class specifies how man or individual gadgets can be constructed. A vector class template is probably declared like this:

Template<class T> class vector
{
T
*
v
;
I
n
t
s
z
;
P
u
b
l
i
c
:
Vector(int);
T
&
o
p

e

r

a

t

o

r

[

]

(

i

n

t

)

;

T

&

e

l

e

m

(

i

n

t

i

)

{return v[i];) // ...

};

The template <class T> prefix specifies that a template is being declared and that an issue T of type *type* can be used inside the declaration. After its advent, T is used precisely like othertype names within the scope of the template assertion. Vectos can then be used like this:

```
Vector<int> vl(20);
Vector<comple
```

x
>
v
2
(
3
0
)
;

Typedef vector<complex> cvec; // make cvec a
synonym for // vector<complex>Cvec v3(40); // v2
and v3 are of the same type vl[3] = 7;
V213] = v3. Elem(4) = complex(7,8) ;

C++ does no longer require the user to explicitly "instantiate" a template; that is, the user want not to specify which variations of a template need to be generated for precise units of template arguments. The motive is that most effective while the program is entire can it be acknowledged what templates need to be instantiated. Many templates might be defined in libraries and lots of instantiations may be without delay and in a roundabout way resulting from customers that don't even understand of the life of those templates. It therefore seemed unreasonable to require the consumer to request instantiations (say, via the use of something like Ada's 'new' operator).

Warding off useless space overheads because of too many instantiations of template functions become considered a primary order, that is, language level hassle instead of an implementation detail. I considered it not going that early (or even past due) implementations would be able to study instantiations of a class for unique template arguments and deduce that everyone or part of the instantiated code could be shared. The solution to this trouble changed into to use the derived class mechanism to make sure code sharing amongst derivedtemplate times.

The template mechanism is absolutely a compile and link time mechanism. No part of the template mechanism needs run-time guide. This leaves the hassle of the way to get the classes and capabilities generated (instantiated) from templates to depend on statistics recognized simplest at run time. The answer turned into, as ever in C++, to apply virtual functions and abstract instructions. Abstract classes used in reference to templates additionally have the effect of presenting better records hiding and higher separation of programs into independently compiled units.

4.10 ANSI and ISO

The initiative to formal (ANSI) standardization of C++ became taken by means of HP at the side of AT&T, DEC, and IBM. Larry Rosler from HP become critical on this initiative. The idea for ANSI standardization turned into written by using Dmitry Lenkov [Lenkov 1989]. Dmitry'sidea cites several reasons for fast standardization of C++:

- C++ is going via a much faster public recognition than most other languages.
- postpone will cause dialects.
- requires a cautious and specific definition imparting full semantics for every languagecharacteristic.

- C++ lacks some crucial functions ... Exception handling, factors of more than oneinheritance, capabilities helping parametric polymorphism, and preferred libraries.

The idea also burdened the want for compatibility with ANSI C. The organizational assembly of the ANSI C++ committee, X3J16, took place in December of 1989 in Washington, D.C. and was attended by means of about forty humans, inclusive of people who took component within the C standardization, individuals who with the aid of now have been "vintage time C++ programmers." Dmitry Lenkov have become its chairman and Jonathan Shopiro have become its editor.

The committee now has more than 250 individuals out of which something like 70 turn up atconferences. The purpose of the committee turned into, and is, a draft widespread for public overview in overdue 1993 or early 1994 with the wish of an official fashionable about two years later. Five that is an ambitious agenda for the standardization of a trendy-cause programming language. To compare, the standardization of C took seven years.

Naturally, standardization of C++ is not simply an American difficulty. From the start, representatives from other nations attended the ANSI C++ meetings; and in Lund, Sweden, inJune of 1991 the ISO C++ committee WG21 changed into convened and the 2 C++ requirements committees determined to maintain joint meetings--starting right now in Lund. Representatives from Canada, Denmark, France, Japan, Sweden, UK, and America have been gift. Substantially, the enormous majority of those country wide representatives were truly long-time C++ programmers. The C++ committee had a tough constitution:

1. The definition of the language need to be unique and comprehensive.
2. C/C++ compatibility needed to be addressed.
3. Extensions past modern-day C++ practice needed to be considered.
4. Libraries had to be taken into consideration.

On top of that, the C++ network was very diverse and definitely unorganized so that the standards committee clearly became an essential focal point of that community. In the quickrun, that is in reality the most important role for the committee.

C compatibility changed into the primary essential debatable issue we needed to face. After some time once in a while heated debate changed into decided that 100 percentage C/C++ compatibility wasn't an option. Neither become extensively reducing C compatibility. C++ was a separate language and no longer a strict superset of ANSI C and couldn't be changed to be this type of superset without breaking the C++ type machine and without breaking millions of strains of C++ code. This selection, often referred to as "As close to C, but no nearer" after a paper written by using Andrew Koenig and me [Koenig 1989a], is the equal that has been reached over and over again by individuals and organizations thinking about C++ and the path of its evolution.

4.11 Retrospective

It is frequently claimed that hindsight is an actual technological know-how. It isn't. The claim is based totally at the false assumptions that we realize all relevant facts about what befell inside the past, that we realize the present-day scenario, and that we have a certainly detached point of view from which to choose the beyond. Usually none of those conditions maintain. This makes a retrospective on something as big, complicated, and dynamic as a programming language in large scale use unsafe.

1. Did C++ be successful at what it changed into designed for?
2. Is C++ a coherent language?
3. What turned into the most important mistake?

Evidently, the replies to these questions are related. The simple solutions are, 'yes,' 'sure,' and 'no longer delivery a bigger library with release 1.0'

4.12 Hopes for the Future

May C++ serve its purpose in community nicely. For that to happen, the language itself shouldbe strong and well specified. The C++ requirements group and the C++ compiler vendors have excellent responsibility here.

Similarly, to the language itself, we need libraries. Absolutely, we need a libraries industry toproduce, distribute, preserve, and teach people. This is emerging. The undertaking is to allow packages to be composed out of libraries from specific vendors. This is hard and can want a few supports from the requirements committee within the form of popular instructions and mechanisms that ease the use of independently evolved libraries.

The language itself, plus the libraries, outline the language that a user de facto writes in. But,only through suitable information of the application regions and design strategies will the language and library features be positioned to correct use. Therefore, there need to be an emphasis on teaching people powerful design techniques and correct programming practices. The various strategies we need have still to be evolved, and most of the satisfactory techniques we do have nonetheless compete with simple lack of information and snake oil. Hope for a long way higher textbook for the C++ language and for programming and design techniques, and specifically for textbooks that emphasize the relationship between language features, accurate programming practices, and exact design techniques.

Techniques, languages, and libraries should be supported by equipment. The days of C++ programming supported with the aid of sincerely a "bare" compiler are nearly over, and the pleasant C++ equipment and environments are beginning to approach the energy and convenience of the nice tools and environments for any language. We will do tons better. Thebest has to come yet, hopefully.

Chapter 5

5.1 Uses of C++

There are numerous benefits of using C++ for developing programs and many applications product based developed in this language only because of its capabilities and safety. Please discover the underneath sections, where makes use of C++ has been widely and efficiently used.

```
            copy_from_user(group_info->blocks[i], grouplist, len))
            return -EFAULT;

        grouplist += NGROUPS_PER_BLOCK;
        count -= cp_count;
    }
    return 0;
}


/* a simple Shell sort */
static void groups_sort(struct group_info *group_info)
{
    int base, max, stride;
```

Below is the listing of the top uses of C++.

- **Applications:** its miles used for the development of new packages of C++. The applications based on the graphic person interface, which are exceedingly used programs like adobe photoshop and others. Many packages of Adobe systems are advanced in C++ likeIllustrator, adobe most advantageous and image geared up and Adobe builders areconsidered as energetic inside the C++ network.
- **Games:** This language is likewise used for developing games. It overrides the complexity of 3D video games. It allows in optimizing the sources. It helps the multiplayer alternative with networking. Makes use of C++ lets in procedural programming for extensive functions of CPU and to offer manage over hardware, and this language is very fast because of which it's far widely utilized in growing specific games or in gaming engines. C++ specially utilized in developing the suites of a sport device.
- **Animation:** there is animated software program, that's evolved with the assist of the C++language. 3D animation, modeling, simulation, rendering software program are referred to as the effective toolset. It's miles broadly used in building real-time, picture processing,cell sensor packages, and visible consequences, modeling which is particularly coded in C++. This evolved software used for animation, environments, motion pix, digital fact, and character creation. Virtual real devices are the maximum popular in these day's amusement international.
- **Web Browser:** This language is used for growing browsers as

nicely. C++ is used for makingGoogle Chrome, and Mozilla internet browser Firefox. A number of the applications are written in C++, from which Chrome browser is one among them and others are like a record machine, the map reduces huge cluster facts processing. Mozilla has other utility additionally written in C++ that is e-mail customer Mozilla Thunderbird. C++ is likewise a rendering engine for the open-source tasks of Google and Mozilla.

- **Database get admission to:** This language is likewise used for developing databasesoftware or open-source database software. The instance for that is mysql, which is one of the most famous database control software and widely utilized in corporations or the various developers. It allows in saving time, cash, commercial enterprise structures, and packaged software. There is different database software get admission to primarily based programs used which can be Wikipedia, Yahoo, youtube, etc. The alternative instance is Bloomberg RDBMS, which facilitates in presenting actual-time economic records to traders. It is in particular written in C++, which makes database get admission to fast and brief or accurate to deliver information concerning commercial enterprise and finance, information round the arena.

- **Compilers:** maximum of the compilers particularly written in C++ language most effective.The compilers which might be used for compiling different languages like C#, Java, and so on. Mainly written in C++ only. It is also used in developing those languages as well as C++is platform-impartial and able to create a diffusion of software.

- **Running systems:** It's also used for growing most of the working structures for Microsoftand few elements of the Apple operating machine. Microsoft windows 95, ninety-eight, 2000, XP, workplace, net Explorer and visible studio, Symbian cell working systems are mainly written in C++ language only.

- **Scanning:** The packages like film scanner or digital camera scanner are also developed inthe C++ language. It's been used for developing PDF era for print documentation, replacing documents, archiving the document and post the files as properly.

Conclusion

C++ is the language that is used anywhere however in particular in systems programming and embedded structures. Here system programming way for developing the running structures or drivers that interface with hardware. Embedded system method matters that are cars, robotics, and appliances. It is having a higher or wealthy network and developers, which helps in the smooth hiring of builders and on line solutions without difficulty. Uses of C++ is known as the safest language due to its security and functions. It is the primary language for any developer to start, who is inquisitive about operating in programming languages. It is straightforward to analyze, as it's far natural idea-based language. Its syntax is quite simple, which makes it clean to write down or develop and mistakes may be effortlessly replicated. Before the use of any other language, programmers preferred to analyze C++ first and then they used different languages. However maximum of the developers try to stay with C++ handiest due to its huge sort of usage and compatibility with more than one structures and software.

CPSIA information can be obtained
at www.ICGtesting.com
Printed in the USA
BVHW061915260321
603510BV00002B/146

9 781802 260601